T0130870

WHY WE STOP GROWING

WHY WE STOP GROWING

GROWING

Etienne Geraert

Zoology Professor

Ghent University, Belgium

ACADEMIA
PRESS

CONTENT

PREFACE

In 2013 Paige Williams asked, "Why do we stop growing?" In her answer she noted, "It's in our genes, but how they exert this control is a mystery".

In the report, which appeared on NBC News, Williams acknowledged that Dr Scott A. Rivkees, a professor of paediatric endocrinology at Yale school of Medicine, had come to the conclusion that we were "just starting to understand the growth-promoting and growth-inhibiting factors of organs".

Earlier findings by Slack (1999), as detailed in his book "On Growth and Form: Spatio-temporal Pattern formation in Biology", led to three very interesting questions associated with growth:

(1) What controls the absolute size of the whole, or why are we bigger than mice? (2) Within a whole, what maintains the constancy of proportions of individual parts? (3) How is a possible change of relative proportions (allometry) produced?

In Geraert (2004) several answers are given in an article titled, "Constant and continuous growth reduction as a possible cause of ageing". In another article called, "A quadratic approach to allometry yields promising results for the study of growth", Geraert (2016) reflects on the mathematics of differential growth. A study on human growth based on data obtained in Belgium from the XIXth century was subsequently added to the literature (Geraert, 2018). These articles together with selected papers written on the growth of nematodes form the basis of this book.

The aim is to show the reader that the growth questions above have been answered. Studies of various animals (worms, arthropods, vertebrates) have showed one and the same phenomenon – and the human being presents no exception. As Needham (1964) already concluded, "It may be taken as established that growth is fundamentally similar in all organisms".

Simple mathematics explain and predict growth – not only growth, but also the entire life cycle.

Some of the working titles of this book included "Ageing starts at birth", "Constant and continuous growth reduction is the cause of ageing", "Growth is never exponential" and "The Laws of Growth".

In general, growth is not well understood. Having studied this complex phenomenon for years, I have come to the conclusion that growth is not time-dependent but relation-dependent. In other words, growth depends on the relationship between several parts of an organism.

I would like to thank Dr G. Packard (Colorado, USA) for mentioning Needham (1957) and the anonymous reviewer of Geraert (2018) for his interesting remarks.

Cited in Needham (1964) "The Growth Process in Animals"

"… our birth is nothing but our death begun …"

(From the poet Edward Young in "Night Thoughts").

INTRODUCTION

Growth occurs equally well in unicellular and multicellular organisms. It is only in multicellular organisms that growth can be considerable; these organisms die, so multicellularity is only temporarily an advantage. The general idea is that growth starts exponentially (Huxley, 1924) until adulthood, and then unicellular offspring are made and only afterwards, ageing starts. Some presume that ageing is caused by the activity of special genes that were not active during growth. If these ageing genes were to be stopped, the adult would remain young and vigorous. Exponential growth means growth without restrictions. If, however, we consider that multicellularity in itself could cause problems, we might as well assume that the restrictions start at the very beginning. When a zygote divides, the many cells become slightly different. Every cell has its own metabolism: some products of each cell positively or negatively influence the metabolism of each other cell. As a consequence, a multicellular organism produces various growth-promoting and growth-inhibiting substances influencing the growth patterns.

To simplify the study of the three-dimensional and complicated growth patterns, it is customary to measure only some distances (or other data) and to compare the measurements. By putting observed values of a growth pattern in an arithmetic graph, curved lines are often found. The nature of the curve has been a subject of controversy for years. Huxley (1924) proposed a power curve. As this curve was not able to explain why growth stops, several other curves and variants have been introduced (see the review in Zeger and Harlow, 1987 for the curves known then). Some of the curves describe growth patterns that differ from what is studied here. For example, West *et al.* (1997, 2001) argued for fractional power laws on the basis that the limiting factor in growth was the formation of branching trees of the circulatory system and that this had an essentially fractal dimensionality.

MATHEMATICAL APPROACHES TO DIFFERENTIAL GROWTH

When growth is studied by comparing the measurements of two body parts over time a curved line is usually observed.

To decide which curve should be used, it has been found helpful to obtain scatter diagrams of transformed variables. To facilitate this, researchers use special graph paper for which one or both scales are calibrated logarithmically, referred to as semi-log or log-log graph paper, respectively.

One can also use arithmetic graph paper and untransformed variables.

The process of curve fitting has the disadvantage that different observers present different curves and equations, so there is a need for a theoretical understanding of growth.

The power curve theory

Huxley's (1924) assumption was that for a theoretical small amount of growth there was a constant ratio between the two growth rates of body part y and body part x.

$$dy/dx = \text{constant k} \qquad (1)$$

This resulted in the formula of allometric growth, first used by Snell (1891)

$$y = bx^k \qquad (2)$$

b and k being constant factors. This formula can also be written

$$\log y = \log b + k \log x \qquad (3)$$

The curvilinear relationship (2) is linearised when the data are plotted on a log-log scale. The slope of that line is represented by the power factor k (known as the allometric coefficient) and log b is the intercept of the line on the y-axis.

The study of growth has been greatly influenced by Huxley's proposal, although doubts have also been expressed. The results on log-log graph paper often show not one single straight line, but two to three consecutive straight lines. These observations have been explained by "sudden changes" in the allometric constant k, which have casted some doubt on the allometric formula (Ford & Horn, 1959 considered them as "methodological artefacts"). Several other formulas have been proposed, a review of which can be found in Zeger & Harlow (1987). The main point is that Huxley's curve did not consider that differential growth was size related.

As Kidwell & Williams (1956) noted: "Huxley (1924) suggested that this equation might express a general law of differential growth. Later (Huxley, 1932), in an attempt to establish a theoretical basis for the equation as a biological law, he derived the formula on the basis of assumptions about growth in general. A number of investigators have postulated different hypotheses to account for the allometric equation as a fundamental biological law of growth, but none has withstood critical analysis. It must be concluded that no satisfactory theoretical basis has yet been found for simple allometry."

Recent publications on that matter are for example, Stern & Emlen (1999), Gayon (2000), Knell *et al.* (2004), Shingleton (2010) and Packard (2012).

The parabola theory

In 1978 and 1979 I published several articles on growth and form in nematodes. I realised that the curves I obtained (when differential growth was studied) fitted best with parabolic curves. No theoretical explanation was given at that time, but I revisited my findings later on in Geraert (2004).

When we compare the growth of one body part in relation to another body part, we have to take into account that the growth of these two body parts is mutually controlled (the more so if the two body parts are functionally related to each other). This means that each additional amount of growth of one body part depends on the additional amount of growth of another body part with which it can be compared.

When we express the additional amount of growth in body part x as Δx and in body part y as Δy, we can translate the slight and gradual changes in shape during growth into slight and gradual changes in Δx and Δy. When these changes have a regular character, as they usually do, we can suppose that a constant Δx provokes a constantly smaller or larger Δy or for each $\Delta x = 1$, $\Delta y2 - \Delta y1 = 2a$ (a = constant factor). The resulting growth relation between x and y will be a quadratic parabola because for this curve "the second differences remain constant" (Batschelet, 1975). This curve can be written as

$$y = a.x^2$$

when the vertex is at the origin, or as

$$y = a.x^2 + b.x + c$$

when the vertex is not at the origin

The values b and c in this formula have a mathematical meaning, not a biological one; they are needed to position the curve in a diagram. The important factor is the quadratic factor a. It can have a positive or negative sign. When the sign is negative it shows that the increase in the y-value decreases and its size shows exactly the degree of change in the growth of body part y relative to x (multiplied by two it gives the exact value of the second growth difference in y for one unit of x). When, during growth, factor a is very low, the curves approach a straight line.

By using this formula, we assume that measurement y is the dependent variable and x the independent variable, but growth in general is more complicated. Moreover, the given measurements are probably not taken along the important growth axes. Never-theless, the negative sign of a in various comparisons indicates why growth stops, namely, because the growth of one body part is constantly and negatively influenced by the growth of another body part. This does not happen suddenly or after some growth has occurred – it starts from the very beginning. Therefore, I hypothesise that when growth starts, a very precise growth pattern ensues that cannot be changed.

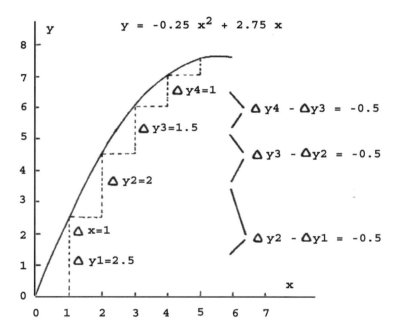

Fig. 1. *When the growth relation between y and x is of the form* $y = -0.25\,x^2 + 2.75\,x$ *then for a constant difference of* $x = 1$ *the second difference in* $y = -0.5$. *This is twice the quadratic factor of* -0.25. *In this theoretical example, factor* $c = 0$ *and so the parabolic curve goes through the origin (courtesy of* Biologisch Jaarboek Dodonaea).

MATHEMATICS OF THE QUADRATIC EQUATION

For the values x_1 and y_1 the quadratic equation reads as follows

$$y_1 = a.x_1{}^2 + b.x_1 + c$$

When one unit is added to x_1, the result is y_2.

$$y_2 = a.(x_1 + 1)^2 + b.(x_1 + 1) + c$$

$$y_2 = a.(x_1{}^2 + 2.x_1 + 1) + b.(x_1 + 1) + c$$

$$y_2 = a.x_1{}^2 + a.2.x_1 + a + b.x_1 + b + c$$

The difference between y_2 and y_1 is Δy_1.

$$\Delta y_1 = y_2 - y_1 = a.x_1{}^2 + a.2.x_1 + a + b.x_1 + b + c - a.x_1{}^2 - b.x_1 - c$$

$$\Delta y_1 = a.2.x_1 + a + b$$

When 2 units are added to x_1, the result is y_3.

$$y_3 = a.(x_1 + 2)^2 + b.(x_1 + 2) + c$$

$$y_3 = a.x_1{}^2 + a.4.x_1 + a.4 + b.x_1 + b.2 + c$$

The difference between y_3 and y_2 is Δy_2.

$$\Delta y_2 = y_3 - y_2 = a.x_1{}^2 + a.4. x_1 + a.4 + b.x_1 + b.2 + c - a.x_1{}^2 - a.2.x_1 - a - b.x_1 - b - c$$

$$\Delta y_2 = a.3 + a.2.x_1 + b$$

The second difference is between Δy_2 and Δy_1.

$$\Delta y_2 - \Delta y_1 = a.3 + a.2.x_1 + b - a.2.x_1 - a - b$$

$$\Delta y_2 - \Delta y_1 = 2.a$$

EARLIER DISCOVERIES

The study of differential growth in the abdomen of the female pea-crab, *Pinnotheres pisum* Leach, brought Needham (1950, 1957) to the conclusion that "For purely empirical purposes a general polynomial relation was fitted to the measurements, but the quadratic proved a very good fit". Later, in his book on growth (Needham, 1964) he interpreted his discovery thus: "For simple comparative purposes, without theoretical implications, the fitting of the best polynomial relation has been advocated, largely because it is mathematically easy to manipulate".

The arithmetic parabola was also used by Kidwell & Howard (1970), Martin (1960) and Walker & Kowalski (1971), whereas Cuzin-Roudy & Laval (1975) adopted the logarithmic parabola for their findings. It is interesting to know that every one of these authors stressed that his discovery was arbitrary and had no biological meaning.

Conclusion

The quadratic equation provides the most suitable answer to questions about several differential growth processes because the curve reflects the result of mutually controlled growth processes. Depending on size and function, mutual control creates a gradual change in shape; the more constant these changes between body parts are, the more the observed measurements will match a calculated parabola.

Huxley's allometric growth formula (the expression of a non-size-related change in shape) was not suitable for the differential growth studies I checked.

THE RESTUDY OF HUXLEY'S MATERIAL

1. The case of *Carcinus maenas*

Huxley & Richards (1931) compared the increase in the width of the abdomen with the increase in carapace length of the shore crab *Carcinus maenas*. The sample was split into three categories: unsexables, females and males. Huxley (1932) gave the measurements for the unsexables and the females using abdomen breadth for y and carapace length for x (both in mm).

I calculated the quadratic equation between both (Fig. 2) thus

$$y = 0.0039 \ x^2 + 0.2685 \ x - 0.467$$

For each 10 mm increase in carapace length (= x) the abdomen breadth (= y) shows a constant secondary increase of 0.78 mm (this is twice the quadratic factor).

Huxley (1932) did not find the single straight line needed to support his theory in the case restudied here. The logarithmic plotting showed a kink in the observations for females as well as for males, so different growth coefficients were observed for younger and older specimens. Huxley (1932) gave several k-values (added on Fig. 3) that he experimentally derived from his figure. Moreover, the constant b was not given. Therefore, it is not possible to compare his (several) equations with the single quadratic equation I obtained. The straight lines found by Huxley (1932) in his log-log diagram (Fig. 3) can be interpreted as mathematical accidents. However, Fig. 2 suggests another three consecutive straight lines, which are also mathematical accidents but on an arithmetic diagram.

Fig. 2. *Comparison of carapace length to abdomen breadth in the shore crab* Carcinus maenas. *The measurements given in Huxley (1932) are represented on a double arithmetic scale (and not on a log-log scale). The open circles are the measurements for the unsexed specimens and for the adult females of the shore crab. The calculated quadratic parabola is added. (courtesy of* Belgian Journal of Zoology).

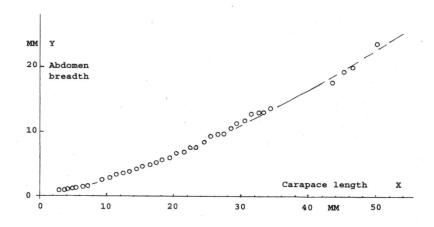

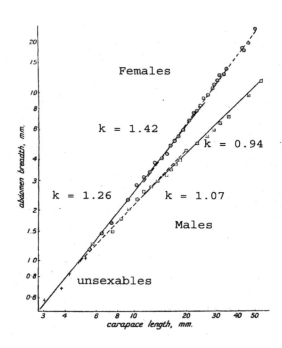

Fig. 3. *Figure taken from Huxley (1932) with the following explanation: "Increase of width of abdomen with increase of carapace length in the shore crab,* Carcinus maenas, *logarithmic plotting". The signs for unsexables, males and females are explained on the graph. The growth coefficients given by Huxley (1932) were also added. (courtesy of* Belgian Journal of Zoology).

Studying the relationship between abdomen breadth and carapace length with a different approach yielded more interesting results (Fig.4), namely, by considering the carapace length (= y) as dependent on the abdomen breadth (= x). The formula is as follows

$$y = -0.0375 \, x^2 + 2.917 \, x + 2.483$$

For each 1 mm increase in the abdomen breadth (= x) there is a constant secondary decrease of 0.075 mm in the carapace length (this is twice the quadratic factor). Therefore, the growth in length of the carapace is negatively influenced by the growth in width of the abdomen.

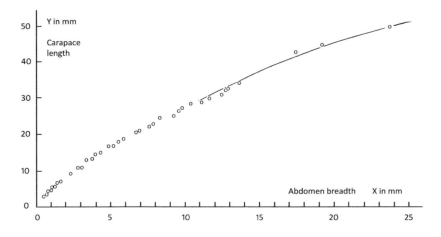

Fig. 4. *Comparison of carapace length and abdomen breadth in the shore crab* Carcinus maenas. *The measurements for the unsexed specimens and for the adult females are given. The calculated quadratic parabola is added.*

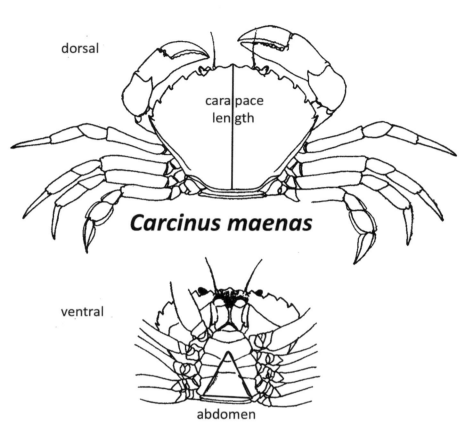

dorsal

carapace
length

Carcinus maenas

ventral

abdomen

Fig. 5. *Dorsal and ventral view of the shore crab with indication of the carapace length and the shape of the abdomen.*

2. The case of *Cyclommatus tarandus*

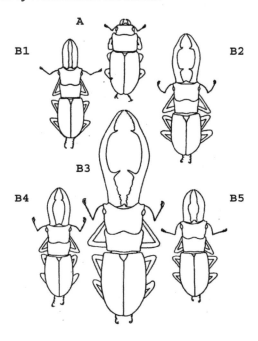

Fig. 6. *Female (A) and five different sized males (B1-B5) of the stag beetle,* Cyclommatus tarandus, *drawn to scale to show the change in form and relative size of the male mandible with the increase of absolute body size.*

Huxley (1927, 1932) studied the growth increase of the mandibles in several species of the Lucanidae, stag beetle. He used the measurements of Dudich (1923) for *Cyclommatus tarandus* (Thunberg, 1806). In this case y = mandible length in mm and x = body length + mandible length also in mm.

I calculated the quadratic equation between both.

$$y = -0.0011 \, x^2 + 0.71 \, x - 11.41$$

For every 10 mm increase in the total length (= x) there is a constant secondary decrease of 0.22 mm in the mandible length (this is twice the quadratic factor). This small quadratic factor indicates that an almost straight line is observed (which may be because measurement y is included in x). The differences between the observed y-values and the calculated y-values are small for whatever x-value is considered (Table I).

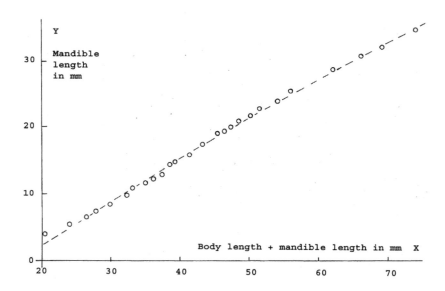

Fig. 7. *Comparison of total length (= body length + mandible length) and mandible length in* Cyclommatus tarandus. *The measurements given in Huxley (1932) are represented on a double arithmetic scale (and not on a log-log scale). The open circles are the measurements for the male specimens. The calculated quadratic parabola is added. (courtesy of* Belgian Journal of Zoology).

Huxley (1932) found that all curves inflected at large absolute sizes. For the smaller animals of *Cyclommatus tarandus* he gave a k-value of 1.97 and a b-value of "just over 0.01" (Fig. 8). Therefore, it is not relevant to compare his curve restricted to the smaller animals with the quadratic one presented here, which includes all the measurements.

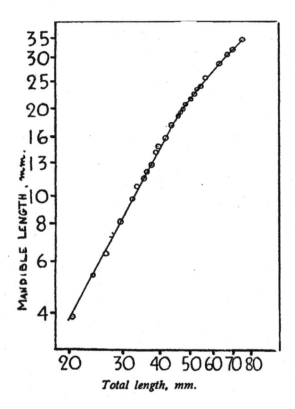

Fig. 8. *Huxley's explanation of his Fig. 35 reads as follows: "Relative growth of male mandibles of stag-beetles (Lucanidae)". I selected only* Cyclommatus tarandus *that is explained as follows: " 'Total length' is true total length. The curve inflects at large absolute sizes; for the remainder of the curve 'k' is nearly 2.0." (courtesy of* Belgian Journal of Zoology).

TABLE I.

Calculated values for Carcinus maenas *and* Cyclommatus tarandus *based on the measurements given in Huxley (1932). The calculated values were obtained by using the quadratic equation shown in the text. The y-values were calculated by a constant increase of the x-values by 10 mm. As a result, the constant second differences in the y-values were obtained. x and y were explained in the text.*

Carcinus maenas				Cyclommatus tarandus			
X-value in mm	Y-value calculated	Increase in Y-values	Second differences	X-value in mm	Y-value calculated	Increase in Y-values	Second differences
10	2.608			20	2.35		
20	6.463	3.855		30	8.9	6.55	
30	11.098	4.635	+ 0.78	40	15.23	6.33	- 0.22
40	16.513	5.415	+ 0.78	50	21.34	6.11	- 0.22
50	22.708	6.195	+ 0.78	60	27.23	5.89	- 0.22

3. The case of *Uca pugnax*

Huxley compared adult male fiddler crabs, *Uca pugnax*. As these adults show a very large variation in the development of the claw, Huxley (1924, 1932) interpreted this also as "growth". An attempt was made to see whether a quadratic parabola can also be used to describe variation in adults.

The quadratic equation I calculated for the males of *Uca pugnax* is as follows.

$$y = 0.097 \ x^2 + 0.42 \ x - 0.045$$

Here, y is the mean mass (in grams) of the large chela and x is the mean mass (in grams) of the rest of the body after removal of the large chela. The calculated values for y are very near to the observed ones, except for the smallest bodies where a slightly different y-value is calculated. The overall similarity is obvious in Fig. 9. The quadratic factor in the above equation is the growth factor. When multiplied by 2, it gives the second difference in the growth of the claw relative to the growth of the rest of the body. For example, for every additional increase of 100 mg of the body, there is a constant increase in the growth of the claw of 1.94 mg (= 2 x 0.97).

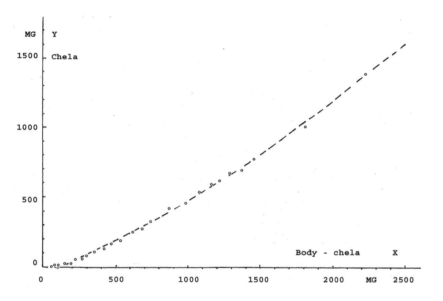

Fig.9. *Diagram representing the measurements obtained by Huxley of the males of the fiddler crab* Uca pugnax. *The dashed line is the calculated quadratic parabola. (courtesy of* Journal of Experimental Biology).

Note: Shih *et al.* (2016) revised the systematics of the family Ocypodidae (Crustacea: Brachyura). The genus *Uca* and the several new genera introduced contain numerous species, most of them unknown in Huxley's time. For this reason, I used the name given by Huxley in this study.

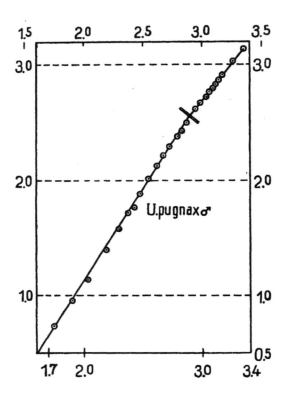

Fig. 10. *Huxley's (1932) explanation of his figure reads as follows: "Increase of the logarithm of absolute chela-weight with the logarithm of body-weight in male fiddler crabs."*

Huxley (1932) realised that the measurements were not on a single straight line as his theory predicted but on two consecutive straight lines. He interpreted this as two "phases" in the growth process. Packard (2012) restudied Huxley's measurements of *Uca pugnax* and presented a two-parameter power function as well as a three-parameter model for describing the observations. The three-parameter model has no biological meaning.

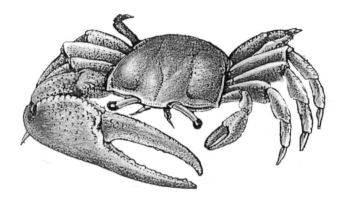

Fig.11. *Frontal view of a male of an* Uca *species showing the large right chela.*

Discussion

Huxley (1924, 1932) based his assumptions mainly on the variation found in adult specimens; the males of the fiddler crab and the stag beetle are good examples. The very large variation observed in these adults was interpreted by Huxley also as "growth". Geraert (2004) followed growth from the new-born to the adult. The restudy of Huxley's material was carried out to see if a quadratic parabola could also be used to describe variation in adults, which is called "static" and "intra-specific" allometry in Gould (1966) and Gayon (2000).

The term allometry was introduced by Huxley & Teissier (1936). It describes the changes in relative dimensions of parts of an organism that are correlated with changes in shape and overall size (Levinton, 1988 in Gayon, 2000).

The allometric equation $y = b.x^k$ needs two factors (b and k) to describe the relationship between the measurements x and y. The meaning of each factor has been debated for many years and was summarised in Gayon (2000), however, no final conclusion was reached.

The quadratic equation has only one factor that shows the constant increase (positive factor) or decrease (minus factor) of one of the measurements for a constant increase in the other measurement with which it is compared.

Huxley (1932) did not find the single straight line on log-log paper needed to support his theory in the three cases restudied here. For the shore crab the logarithmic plotting showed a kink in the observations for females as well as for males, so a different growth coefficient was observed for young females (males) and older females (males). Huxley (1932) gave several k-values (added on Fig. 3) that he experimentally derived from his figure. Moreover, the constant b was not given. Therefore, it is not possible to compare his (several) equations with the single quadratic equation obtained. The straight lines found by Huxley (1932) in his log-log diagram (Fig. 3) can be interpreted as mathematical accidents.

For the stag beetles, Huxley (1932) found that all "curves" inflected at large absolute sizes, whereas for the smaller animals of *Cyclommatus tarandus* he gave a k-value of "just over 0.01" (Fig. 8). Therefore, it is not relevant to compare his curve restricted to the smaller animals with the quadratic one presented here (Fig. 7), which includes all the measurements.

A similar result was found for the comparative study of the male fiddler crab; in this case weight was compared, not length. Here again, Huxley (1932) needed two curves (Fig.10) instead of one for the quadratic relationship (Fig. 9).

Huxley continued propagating the power curve for describing growth. His proposal has been generally accepted, including, for example, in recent times by Knell *et al.* (2004) for the stag beetle, and Shingleton (2010) and Packard (2012) for the fiddler crab.

The quadratic curve closely follows all the observations, including those relating to the smallest and the largest animals; one single quadratic factor explains the entire process.

Champy (1924), cited by Gayon (2000), argued that the relative growth process was adequately described by a parabolic curve of the shape $V = at^2$, which is a special case and is different from the curve proposed here. Teissier (1931), cited by Gayon (2000), observed that Champy's law was indeed a good approximation for some insects. Using curve-fitting programs, Martin (1960), Kidwell & Howard (1970) and Walker & Kowalski (1971) observed that a parabolic curve gave the best approximation for their measurements on growth; every one of these authors stressed that his discovery was arbitrary and had no biological meaning.

Conclusion

The quadratic equation explains intra-specific allometry in a satisfactory way.

Huxley's proposal to use a power curve to explain intra-specific allometry is no longer applicable. By using the quadratic curve there is no need for such explanations as "sudden changes in the growth factor" or "the curve inflects at large absolute sizes". The quadratic factor of the quadratic equation is the only factor that describes relative growth.

GROWTH IN NEMATODES

In animal-parasitic nematodes the pharynx can change shape from a rhabditid-bulbous form to a strongylid-cylindrical type. Other animal parasites and several plant-parasitic nematodes grow substantially in width during their development. Such drastic changes do not occur in free-living nematodes. Through a comparative study of the post-embryonic development of nematodes belonging to several orders an attempt is made to explain these changes.

Despite the structural diversity of the pharynx, we can broadly recognise three main types (Decraemer, 2016):

1. The one-part type is either entirely cylindrical or slightly enlarged anteriorly/posteriorly. It is the simplest type of pharynx and is typical of the free-living enoplids and mononchs.

2. The flask-shaped type consists of a corpus and a swollen muscular, glandular postcorpus, typical of the free-living dorylaims.

3. The tripartite pharynx is the most complex and has three well-defined main parts: a corpus, a slender part (isthmus) and a terminal bulb. The corpus may be further subdivided into pro- and metacorpus, and the metacorpus may contain a valve. A complex pharynx type is found in various rhabditids, diplogasterids, tylenchs and aphelenchs, showing differences in the degree of development of the metacorpus. In tylenchs a non-muscular isthmus and three pharyngeal glands follow the corpus.

In Fig. 12 three forms are depicted: the tylenchiform pharynx (Fig. 12 A), the rhabditiform pharynx (Fig. 12 B) and the flask-shaped form (Fig. 12 C).

Fig. 12. *Light microscopic views of three pharynges: A.* Tylenchorhynchus leviterminalis *as an example of a tylenchiform pharynx; B.* Acromoldavicus skrjabini *as an example of a rhabditiform pharynx; C.* Eudorylaimus pseudocarteri *as an example of a flask-shaped pharynx (A from Troccoli & Geraert, 1995*; *B from Karegar* et al., 1997; *C from Choi* et al., *1997).*

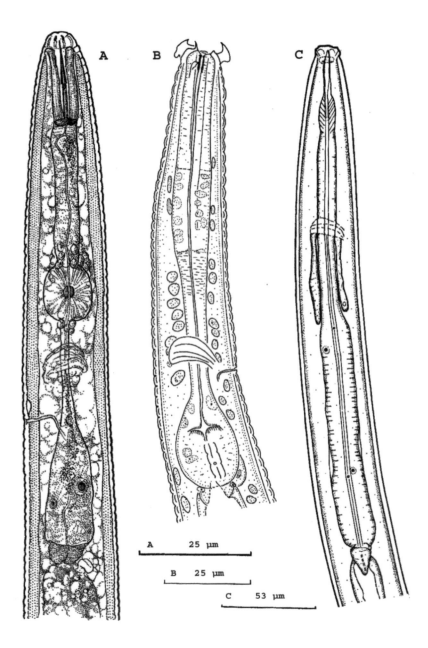

A	25 µm
B	25 µm
C	53 µm

Growth in nematodes with a tylenchiform pharynx

The tylenchiform pharynx occurs in tylenchs and aphelenchs. Both groups are characterised by the presence of a stylet with which they penetrate a host to suck the content. Most of them are plant parasites, but some are animal parasites (e.g. *Seinura*). Their pharynx is characterised by a muscular median bulb followed by a thin isthmus and a non-muscular region containing three pharyngeal glands. These glands can either constitute a terminal bulb, offset from the intestine (Fig. 12 A) or lie free in the body cavity alongside the intestine.

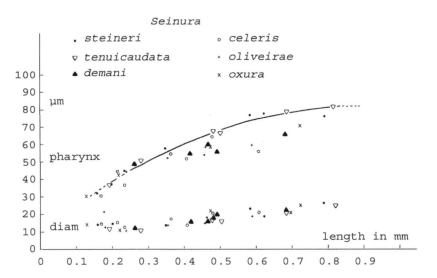

Fig. 13. *Pharyngeal length and body width compared to body length in six* Seinura *species. These species have the shortest pharynx of the nematodes. (courtesy of* Nematologica).

Seinura contains species with a short pharynx – the intestine starts shortly posteriorly to the median bulb and the three glands lie alongside the intestine. The pharyngeal growth pattern is more or less comparable among the several species and the final pharyngeal length is almost or already reached by the last juvenile stage. The calculated parabola for *S. tenuicaudata* is as follows (L = body length).

$$\text{pharynx length} = -0.09138\ L^2 + 0.16155\ L + 0.01026$$

Very different relations between body width and body length (= L) have been found in adult tylenchs that show a vermiform shape (all measurements in mm). Compare, for example, *Helicotylenchus vulgaris* (1) with *Hemicycliophora arenaria* (2).

$$\text{width} = -0.00914\ L^2 + 0.04425\ L + 0.001193 \quad (1)$$

$$\text{width} = 0.0265\ L^2 + 0.01544\ L + 0.00969 \quad (2)$$

In several plant-parasitic tylenchs, the mature females become thicker than usual, ranging from the obese *Anguina* to the cyst-forming *Heterodera*.

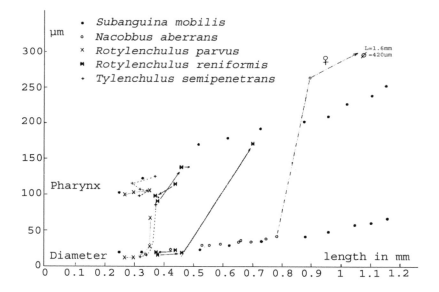

Fig. 14. *Pharynx length and body width compared to body length in some obese and saccate tylenchs. (courtesy of* Nematologica).

In *Subanguina mobilis* the pharynx grows considerably and almost constantly during the whole developmental process. The shortening of the body in *Rotylenchulus reniformis* from the second to the third juvenile stage goes together with a shortening of the pharynx and vice versa. The lengthening of the body from the fourth juvenile stage to the immature female corresponds with a lengthening of the pharynx. *Rotylenchulus parvus* shows a similar picture. A different growth pattern is observed in *Tylenchulus semipenetrans*. The pharynx grows slowly but constantly although the body length shortens from the third juvenile stage to the immature female. This shrinkage is typically restricted to the posterior region so that the original positions of the genital primordium and the excretory pore in the second juvenile stage (at 54%) do not correspond to the positions of the vulva and excretory pore in the young female (at 80%).

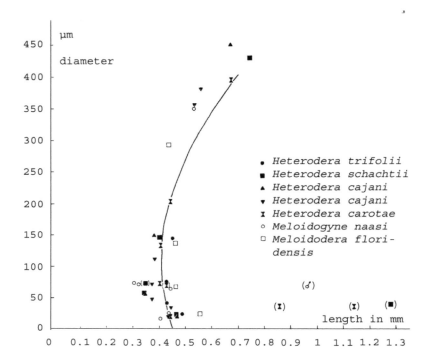

Fig. 15. *Relation between body length and body width in some saccate tylenchs. (courtesy of* Nematologica).

The thickening of these obese to saccate plant parasites can be slight but constant, as for *Subanguina;* or it can only occur in the immature, vermiform female (*Rotylenchulus, Tylenchulus, Nacobbus);* or it can occur from the moment the second stage juvenile has found a suitable host (*Heterodera, Meloidogyne, Meloidodera).* The measurements of *Heterodera, Meloidogyne* and *Meloidodera* show that the first thickening of the body in females corresponds with a shortening of the body. The parabola of the shape calculated for *H. carotae* (both measurements in mm) is formulated thus

$$L = 3.4975 \text{ width}^2 + 0.7359 \text{ width} + 0.44886$$

In the female of *H. carotae* the pharynx and the body length grow only slightly, whereas in the male both measurements become distinctly larger. The growth of the female body volume corresponds, however, with a considerable growth in the volume of the median bulb. In the male the lengthening of the pharynx corresponds with a diminution of the volume of the median bulb.

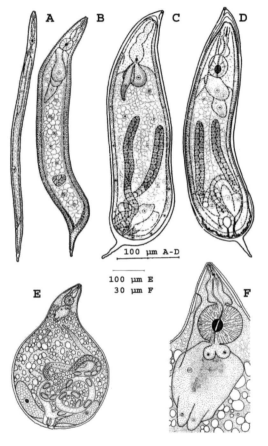

Fig. 16. *The post-embryonic development of* Meloidogyne, *a tylench that does not grow in length but in width; second juvenile stage (A, B); fourth stage female juvenile (C); adult female shortly after the fourth molt (D); fully developed female (E); an enlargement of the anterior region (F). (A-D after Triantaphyllou & Hirschmann, 1960; E-F after Eisenback, 1989)*

DISCUSSION

Pharyngeal growth

During post-embryonic development the growth of the pharynx follows the growth of the body. This general statement can best be emphasised by two examples: (1) In *Rotylenchoides reniformis* the shortening of the body results in the shortening of the pharynx; (2) In *Heterodera carotae* the juveniles becoming females grow only slightly in length and so does the pharynx. The juveniles becoming males grow considerably in length and so does the pharynx.

It was found that when there is no lengthening of the pharynx during development, maturing females overcome the increasing demand for food through two solutions: (1) in the saccate females the median bulb enlarges (Fig. 16 F); (2) in others the time spent feeding increases.

Body width

The findings give rise to two questions about the female nematode: firstly, how does a female nematode search for food, and secondly, how does it achieve the production of as many eggs as necessary or possible?

One group of nematodes fulfils these basic requirements by developing an appropriate way of feeding at the same spot so that they lose their vermiform shape, either from the hatching stage, like *Heterodera* and *Meloidogyne*, or from the adult stage, as in the case of other saccate tylenchs.

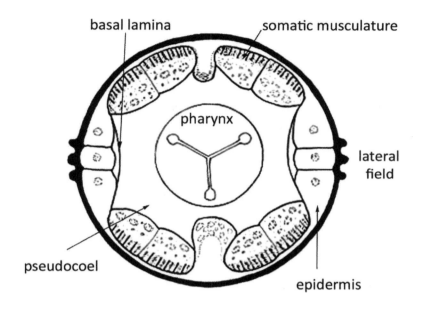

Fig. 17. *Diagrammatic presentation of a transverse section at the level of the posterior pharynx region of a tylench with a lateral field consisting of three ridges.*

Another group have evolved a cuticle with large telescopic annuli that during maturation permit more body elongation than body widening, so that older females become relatively thinner. The majority of the "ring nematodes" (Criconematidae) belong to this second group.

A third group have developed a lateral field that is used as storage for additional cuticular material that is immediately available when the eggs are ripening and the female is widening. The processes of ripening and widening are intimately linked as egg transport and egg laying are only made possible by the kneading action of the body wall muscles.

Growth of the cylindrical pharynx in free-living nematodes

For the relation between pharynx length and body length the data obtained for the six species in Fig. 18 form almost a straight line from the small *Tylencholaimus* to the large *Labronema* (although for most species a slight slowing down occurs). The formula for *Labronema thornei* is as follows (L = body length).

$$\text{pharynx} = -0.01827 \ L^2 + 0.28388 \ L + 0.00942$$

From the study of body width in relation to body length it can be concluded that the species studied become relatively thinner during growth, with the body width 2-3% of the body length. The formula for *Labronema* is as follows

$$\text{width} = -0.00008 \ L^2 + 0.02232 \ L + 0.00669$$

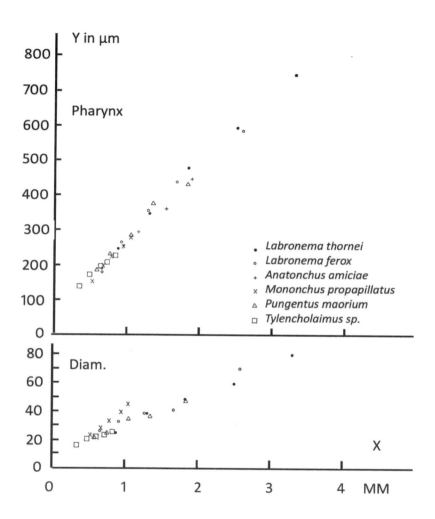

Fig. 18. *Comparison between body diameter, pharynx length and body length in some free-living nematodes with a cylindrical pharynx (courtesy of* Nematologica).

Growth of the rhabditiform pharynx in free-living nematodes

Panagrellus redivivus was studied twice. The curve fitted to the observations for pharyngeal length in relation to body length during post-embryonic growth proved to be a quadratic parabola. During adulthood the pharyngeal length remains almost unchanged.

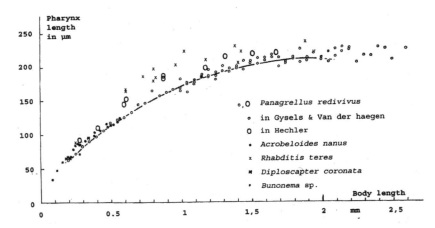

Fig. 19. *Relation between body length and pharynx length in several free-living rhabditids. A calculated arithmetic parabolic curve has been added for the measurements found by Gysels & van der Haegen (courtesy of* Nematologica).

The growth of the pharynx in relation to the body length is apparently very stable. From the start, the animals studied by Hechler (1970) had a slightly longer pharynx (or a slightly shorter body) than the specimens examined by Gysels & van der Haegen (1962), and that difference remains throughout development. Both series of results form a smooth, curved line during development, creating a plateau when the adult stage is reached.

The calculated parabolic curve gives a very close approximation for these observations.

Formula for Hechler's measurements (L = body length in mm):

pharynx length (in μm) = -0.00008418 L² + 0.25854 L + 23.24

Formula for Gysels & van der Haegen's measurements:

pharynx length = -0.00005322 L² + 0.2008 + 24.37

For every 100 μm growth in body length the additional growth of the pharynx shows a constant decrease of 1.6836 μm (= second difference) for Hechler's results and 1.0644 μm for the results found by Gysels & van der Haegen. Several other rhabditids were also studied and their results are added to the graph. They show a similar growth pattern to that of the *Panagrellus*.

Growth of body width in free-living nematodes with a rhabditiform pharynx

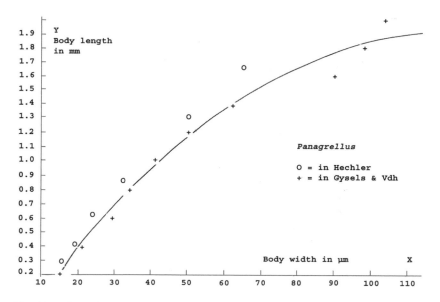

Fig. 20. *Relation between body length and body width in free-living rhabditids. A calculated arithmetic parabolic curve has been added for the measurements found by Gysels & van der Haegen.*

In this case the relation between length and width follows a quadratic curve in which the body length depends on the body width. The mean values for body width given by Hechler follow the quadratic parabola calculated for Gysels & van der Haegen very closely. The measurements of the latter show a slightly wider body in the juveniles, and the adult females are much wider. Because of the larger variability the calculated curve does not closely match the observations.

The calculated curve for Gysels & van der Haegen's measurements is as follows (L = body length; W = body width):

$$L = -150 \ W^2 + 36.3 \ W - 0.283 \text{ (both measurements in mm) or}$$
$$L = -0.15 \ W^2 + 36.3 \ W - 283 \text{ (both measurements in } \mu m)$$

Therefore, for every 10 μm additional growth in body diameter, there is a constant decrease in body length of about 30 μm (this is twice the quadratic factor). A rather uniform picture is obtained for the relation between body length and body width in this free-living group.

Pharyngeal growth in animal-parasitic nematodes with a rhabditiform pharynx

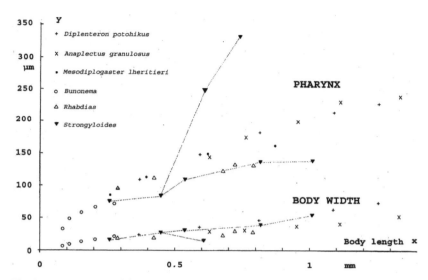

Fig. 21. *Comparison of the measurements of some free-living rhabditids (*Diplenteron, Mesodiplogaster, Bunonema*), one free-living araeolaimid (*Anaplectus*), the free-living generation of some animal-parasitic rhabditids (*Rhabdias, Strongyloides*) and the parasitic generation of* Strongyloides. (*courtesy of* Nematologica).

There are both free-living and infective generations of the *Strongyloides* species. The free-living generation retains the rhabditid pharynx of the hatched juveniles, whereas the infectives lose the typical valvulated bulb. The rhabditid pharynx in the second stage free-living juvenile measures 80-90 μm long, whereas in the infective juveniles the pharynx elongates to 200-260 μm. Nematodes of the infective generation become much longer than those of the free-living generation (Basir, 1950; Matoff, 1936).

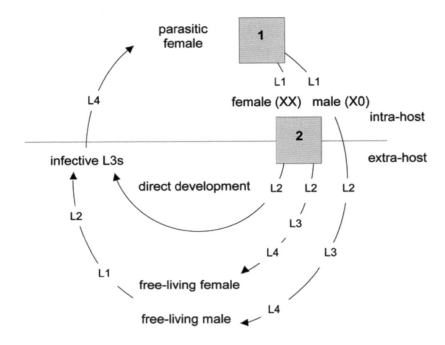

Fig. 22. *The life cycle of* Strongyloides. *L1-L4 are the several juvenile stages.*

Similarly, the newly hatched juveniles of *Rhabdias fuscovenosa* develop either into a free-living generation or an infective generation. The free-living generation retains the rhabditid pharynx of the hatched juveniles, whereas the infectives lose the typical valvulated bulb. Nematodes of the infective generation become much longer than those of the free-living generation. However, in both generations the animals with a similar body length also have a similar pharyngeal length and a similar body width (Chu, 1936).

Throughout its development *Syphacia* shows a rhabditid pharynx with a valvulated terminal bulb and a thick body (Chan, 1952).

Two *Cosmocerca* species are typified by a slightly different pattern. Following the free-living stages, the *C. ornata* develops into a slender juvenile with a pharynx featuring a poorly developed terminal bulb. After entering the host, the parasite doubles in width, the pharynx elongates and the terminal bulb enlarges. In *C. longicauda* this body widening and pharynx development occurs only when the animal reaches its final location within the host (after travelling from lung to large intestine).

Some strongylids are also identified by a change from a rhabditid pharynx in the younger stages to a cylindrical pharynx in the later and adult stages. The change in pharyngeal shape results in a lengthening of the pharynx as can be seen for *Oesophagostomum venulosum* in Fig. 23 (Goldberg, 1951), which also gives the measurements for the stages of *Trichostrongylus axei* with a strongylid pharynx (from the parasitic third juvenile stage to the adult female) (Douvres, 1957).

For some ascarids and strongylids, no stage has been identified as featuring a more or less rhabditid shaped pharynx; the pharynx is always cylindrical usually with a distinct division into an anterior muscular part and a posterior glandular part. The results for the growth of the pharynx for an *Oncophora* species are added to Fig. 23 (described as *Camallanus sweeti* by Moorthy, 1938 but synonymised with *Oncophora* by Rigby & Rigby, 2014).

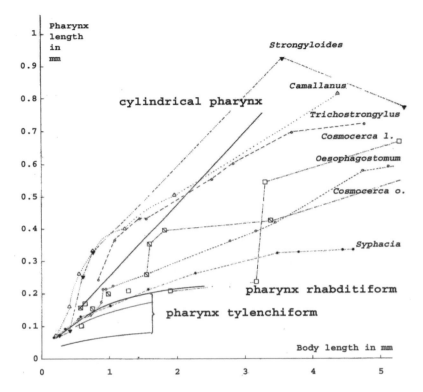

Fig. 23. *Comparison of relative growth in the pharynges of the nematodes; unbroken lines refer to free-living or plant-parasitic forms, broken lines to animal-parasitic species. The lowest unbroken line refers to the line obtained in Fig. 13; the bracket indicates the area in which pharynges of tylenchs are found; the line representing the rhabditiform pharynx is the line obtained in Fig. 19; the line of the cylindrical pharynx comes from Fig. 18. (courtesy of* Nematologica).

Discussion

One of the results of this comparison shows that relative growth is not a haphazard phenomenon. This is especially obvious for pharyngeal growth in free-living rhabditids. It is interesting to note that all the observations of several animals by different authors follow the same curve, and particularly to note that the difference of a few microns in the second stage juvenile of *Panagrellus redivivus* (observed in the two studies and depicted in Fig. 19) continues to exist throughout the nematode's development and into the adult stage.

All the observations show that the growth of the pharynx in relation to the growth in body length follows a quadratic curve. Comparing the results for the pharyngeal growth in plant-parasitic and free-living nematodes (full lines in Fig. 23), we can observe that the tylenchs have the shortest pharynx (usually characterised by the presence of a median, muscular bulb); the slightly longer pharynx of the rhabditids is characterised by a terminal, valvular muscular-glandular bulb; an entirely cylindrical pharynx is typical for the mononchs; but some dorylaims (with a pharynx divided into an anterior, slender, muscular part and a posterior, muscular-glandular part) are included on the same line.

Based on this comparison we can conclude that tylenchs and rhabditids can feed efficiently with a much shorter pharynx than mononchs or dorylaims of the same body length, and that during growth this difference becomes more and more pronounced. The presence of a muscular bulb is an advantage as a smaller portion of the animal is occupied by the pharynx. A distinct pharynx structure could also be related to a different feeding regime. However, in plant-parasitic and free-living forms these types of pharynges show a limited growth.

One of the aims of this chapter is to show that for a pharynx with a median or terminal bulb the limiting factor for pharyngeal growth is body width, which together with body length, determines body volume. A large body can produce many eggs – highly specialised parasites often produce very many eggs to ensure that some of their offspring find a suitable host; therefore, they need a large body volume. A large body volume can be caused by an excessive growth in width, producing short, thick females; an excessive growth in length, producing long, thin females; or an equal growth in length and width, so that the body proportions remain about the same during development.

The observations suggest that as soon as a nematode becomes stouter than it was at the youngest stage, it loses some motility. For optimum movement, the animal must become relatively thinner during development or its original vermiform shape must be maintained.

By contrast, the growth of a pharyngeal bulb necessarily means a growth in width of this structure to maintain its shape. Therefore, animals with a bulbous pharynx have to meet two contradicting requirements during their development: to remain active they must become relatively thinner, but to ingest their food they must become as stout as the bulb width requires. In the larger free-living rhabditids the pharynx reaches a maximum length of 200-250 μm. However, body widening beyond a certain point creates difficulties in egg laying. The eggs develop inside females that become too wide, resulting in death of the female.

Those animal parasites that need a much larger body volume than the free-living rhabditids overcome the structural difficulties in two ways. First, they can change the pharyngeal shape. In this most spectacular phenomenon the loss of the efficient bulb is overcome by an obvious lengthening of the pharynx (Fig. 23: *Oesophagostomum*). Second, a change in width can occur but only when the animal has reached its final environment and can lose part of its motility.

To develop a rhabditid pharynx, a female nematode has to be wider than 1/20-1/25 of its length (4-5%), otherwise it must have a cylindrical pharynx. A cylindrical pharynx is mainly found in the longer nematodes (above 25 mm only cylindrical pharynges are found) and a rhabditid pharynx mainly features in the shorter animals.

Conclusion

During post-embryonic growth body length, pharynx length and body width are interdependent, but their dependence cannot be seen when these measurements are expressed as simple ratios (ratio β = body length/pharynx length and ratio α = body length/body width) because when plotted on a graph, they do not follow a straight line through the origin. However, the dependence is very clear when a quadratic curve is calculated.

The newly hatched juveniles of each nematode species have typical length to width ratios that result in their being either stout or thin while still retaining their vermiform shape. During post-embryonic development the relationship between length and width changes according to the nematodes' behaviour. When the animals remain active, they retain their vermiform shape, and an increase in body volume is the consequence of a growth in length rather than a growth in width. When the animal becomes sedentary, they can lose their vermiform shape and with that the capability to move.

Each type of pharynx has its own growth pattern, but two major trends can be distinguished. First, a muscular, valvular bulb makes a pharynx much shorter and apparently limits the growth in length of the body; animals with such a bulb can grow in girth when they do not need to move around. Second, a more cylindrical pharynx is much longer than a bulbous pharynx and allows almost unlimited growth in length. Animal parasites that have juveniles with a rhabditiform pharynx may transform this pharyngeal shape into a more cylindrical form because otherwise they may not reach the necessary body length to body width ratio for egg production and movement. Free-living nematodes usually do not change much in pharyngeal shape or body shape. The plant-parasitic tylenchs also do not show a change in pharyngeal shape, but they can increase remarkably in girth to such an extent that they become sub-spherical, e.g. *Globodera, Heterodera* and *Meloidogyne (*Fig. 16). In some insect-parasitic tylenchs the parasitic generation has a much longer pharynx. Animal-parasitic nematodes can retain their rhabditiform pharynx either by a gradual increase in girth (*Enterobius*) or by a sudden increase (*Cosmocerca*). In both cases body width reaches at least 4% of body length. Animal-parasitic nematodes can change in form from those with a juvenile rhabditiform pharynx into those with a cylindrical pharynx or they can retain the juvenile cylindrical pharynx. Insect- and animal-parasitic nematodes are usually much thinner (<4% of body length) and can grow much longer than plant-parasitic tylenchs.

Growth of the rat skull

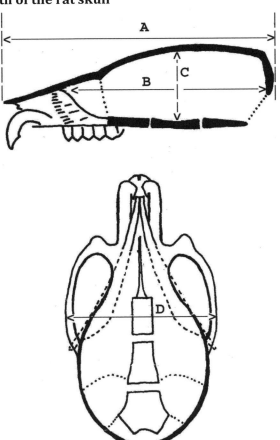

Fig. 24. *Diagram of a rat skull seen laterally (top) and from above (bottom), indicating some of the measurements that were made. A = total skull length; B = endocranial length; C = endocranial height; D = bizygomatic width.*

Ford and Horn (1959) studied the growth of the skull of 40 albino rats of the Wistar strain. Groups of five rats were killed at birth, at ten, fifteen, twenty, thirty, forty, sixty and eighty days of age. Immediately after death radiographs were taken of each rat and various measurements to 0.1 mm were recorded.

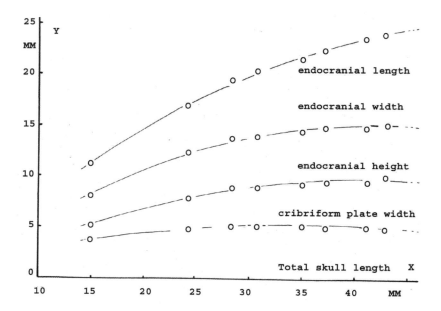

Fig. 25. *Rat skull: four different measurements compared to the skull's total length. The corresponding parabolic curves are added. (courtesy of* International Journal of Developmental Biology).

I obtained the following formulas (x is always the total skull length).

Endocranial length: $y = -0.00925 x^2 + 0.993 x + 1.431$

For each mm growth in total skull length the additional growth in endocranial length diminishes by 0.0185 mm (= twice the quadratic factor)

Endocranial width: $y = -0.0081 x^2 + 0.701 x + 0.008$

For each mm growth in total skull length the additional growth in endocranial width diminishes by 0.0162 mm (= twice the quadratic factor)

Endocranial height: $y = -0.00637 x^2 + 0.527 x + 1.171$

For each mm growth in total skull length the additional growth in endocranial height diminishes by 0.01274 mm (= twice the quadratic factor)

Cribriform plate: $y = -0.00379 x^2 + 0.262 x + 0.711$

For each mm growth in total skull length the additional growth in the cribriform plate diminishes by 0.00758 mm (= twice the quadratic factor).

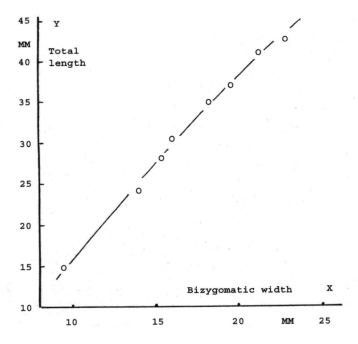

Fig. 26. *Rat skull: total length compared to bizygomatic width. The calculated parabolic curve is added. (courtesy of* . International Journal of Developmental Biology).

The growth in length of the skull negatively influences the growth of the four measurements tested above, each of which show a progressive reduction in growth. The growth pattern of the cribriform plate presents the most interesting finding because the width diminishes long before adulthood is reached. The calculated quadratic factors are important – when multiplied by two they give the growth reduction in mm of that distance for each mm of growth in length of the skull. For two measurements (endocranial width and height) the calculated curves suggest that a maximum is reached.

In Fig. 26 the greatest width in the rat skull (the bizygomatic width) is compared to the length of the same skull. Also here, the calculated curve is added showing that the growth in length of the skull is negatively influenced by the width.

The following formula gives the calculated curve.

Skull length: $\qquad y = -0.02343\,x^2 + 2.9\,x + 10.661$

For each mm growth of the bizygomatic width ($= x$) the additional growth in skull length diminishes by 0.04686 mm (= twice the quadratic factor)

TABLE II

The dimensions of the rat skulls as published by Ford & Horn (1959). All measurements are in millimetres.

AGE IN DAYS	TOTAL SKULL LENGTH	ENDO-CRANIAL LENGTH	ENDO-CRANIAL WIDTH	ENDO-CRANIAL HEIGHT	CRIBRIFORM PLATE WIDTH	BIZYGO-MATIC WIDTH
0	14.78	11.22	8.12	5.16	3.72	9.36
10	24.20	17.04	12.38	7.88	4.92	14.02
15	28.06	19.36	13.71	8.92	5.10	15.36
20	30.54	20.50	13.91	8.92	5.16	16.10
30	34.76	21.64	14.34	9.22	5.18	18.18
40	37.02	22.50	14.85	9.50	5.14	19.48
60	40.88	23.82	15.05	9.62	5.12	21.26
80	42.74	24.16	15.16	9.94	5.02	22.86

Growth in a chicken

Kidwell and Williams (1956) studied a chicken variety (Dark Cornish fowl) present in Louisiana State University. Eighty-three males and 101 females were observed over ten weeks. Six measurements (to the nearest mm) on each chicken were taken at 1 week of age and at weekly intervals thereafter. Weight was recorded to the nearest gram.

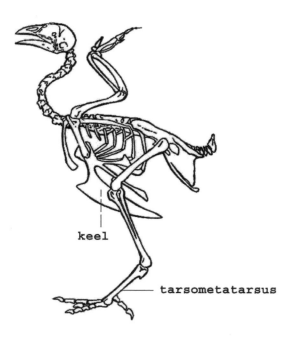

Fig. 27. *Skeletal anatomy of a chicken, indicating the two bones used in this study.*

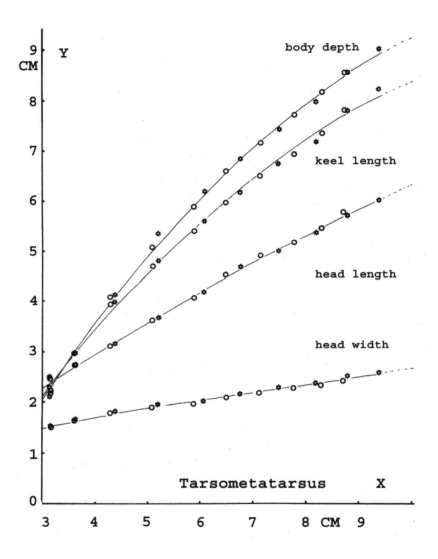

Fig. 28. *Dark Cornish fowl: four different measurements compared to the tarsometa-tarsus length. Females are represented by circles, males by stars. The parabolic curves are added. (courtesy of* . International Journal of Developmental Biology).

Fig. 28 shows the growth of the tarsometatarsus compared to the growth of various body measurements. For three measurements the results show a similar length at 1 week of age but a different increase during the following weeks. All measurements closely follow the quadratic parabola added to the figure.

The following formulas were obtained (x = tarsometatarsus length).

$$\text{Body depth:} \quad y = -0.07733\ x^2 + 2.026\ x - 3.33$$

For each mm growth in tarsometatarsus length the additional growth in body depth diminishes by 0.15466 mm (= twice the quadratic factor)

$$\text{Keel length:} \quad y = -0.06236\ x^2 + 1.69\ x - 2.34$$

For each mm growth in tarsometatarsus length the additional growth in keel length diminishes by 0.12472 mm (= twice the quadratic factor)

$$\text{Head length:} \quad y = -0.00836\ x^2 + 0.6835\ x + 0.3724$$

For each mm growth in tarsometatarsus length the additional growth in head length diminishes by 0.01672 mm (= twice the quadratic factor)

$$\text{Head width:} \quad y = -0.00391\ x^2 + 0.2083\ x + 0.9241$$

For each mm growth in tarsometatarsus length the additional growth in head width diminishes by 0.00782 mm (= twice the quadratic factor)

TABLE III

Comparison of measured and calculated values for body depth in relation to tarso-metatarsus length in Dark Cornish fowl. The measurements for females and males are mixed. In the first column the observed values are given together with the units in brackets. In the second column the corresponding observed values are given. In the third column the calculated Y-values are given for the corresponding X-value. In the fourth column the calculated Y-values for the units of X are given. Column five shows the differences between the calculated values per cm tarsometatarsus increase. The last column shows the constant decrease of the increase.

Tarsometatarsus = x	Body depth = y				
Measured	measured	calculated			
		Per measured x-value	Per unit of x	Differences per unit of x	Second difference
3.16	2.29	2.30			
3.17	2.32	2.32			
3.62	2.99	2.99			
3.64	2.98	3.02			
(4)			3.53672		
4.31	3.93	3.97		1.33003	
4.4	3.96	4.09			- 0.15466
(5)			4.86675		
5.1	5.12	4.99		1.17537	
5.24	5.36	5.16			- 0.15466
5.89	5.91	5.92			
(6)			6.04212		
6.1	6.2	6.15		1.02071	
6.49	6.61	6.56			- 0.15466
6.77	6.85	6.84			
(7)			7.06283		
7.15	7.18	7.20		0.86605	
7.51	7.41	7.53			- 0.15466
7.78	7.67	7.75			
(8)			7.92888		
8.22	7.94	8.10		0.71139	
8.28	8.14	8.15			- 0.15466
8.74	8.52	8.47			
8.8	8.52	8.51			
(9)			8.64027		
9.39	9	8.88			

Fig. 29 shows the dependence of the tarsometatarsus length on body weight. Also here, the observed values match the calculated curve. The extrapolation of the curve suggests that during the observations the tarsometatarsus reached almost its maximum value. This relationship seems logical as the tarsometatarsus bone has to support the weight of the chicken. It is possible that a chicken heavier than 1.4 kg cannot be supported by the tarsometatarsus and broken legs are then observed.

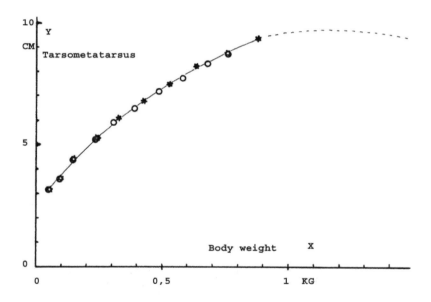

Fig. 29. *Dark Cornish fowl: tarsometatarsus length compared to total weight. Females are represented by circles, males by stars. The extrapolation of the curve shows that the tarsometatarsus is theoretically almost at its maximum length. (courtesy of* . International Journal of Developmental Biology).

The following formula was obtained where y = tarsometatarsus in cm and x = weight in kg.

$$y = -4.922 \, x^2 + 11.95 \, x + 2.537$$

For every 100 g of additional growth in body weight the additional growth in tarso-metatarsus length diminishes by 0.09844 cm (= twice the quadratic factor)

TABLE IV

Original Table from Kidwell & Williams (1957).

Week by Sex Means of Six Body Measurements Taken on 83 Male and 101 Female Dark Cornish Chicks.

Week	Sex	Weight gm	Keel Length cm	Tarso-metatarsus cm	Body Depth cm	Head Width cm	Head Length cm
1	Male	56.42	2.12	3.16	2.29	1.52	2.50
1	Female	55.31	2.21	3.17	2.32	1.49	2.49
2	Male	99.42	2.99	3.64	2.98	1.66	2.74
2	Female	97.69	3.09	3.62	2.99	1.65	2.75
3	Male	155.25	4.13	4.40	3.96	1.80	3.14
3	Female	152.37	4.09	4.31	3.93	1.77	3.13
4	Male	249.13	4.83	5.24	5.36	1.95	3.66
4	Female	236.41	4.72	5.10	5.12	1.88	3.63
5	Male	329.04	5.62	6.10	6.20	2.03	4.20
5	Female	309.64	5.42	5.89	5.91	1.97	4.06
6	Male	430.07	6.20	6.77	6.85	2.17	4.69
6	Female	395.82	5.96	6.49	6.61	2.11	4.55
7	Male	530.06	6.76	7.51	7.41	2.29	5.04
7	Female	487.51	6.49	7.15	7.18	2.20	4.88
8	Male	639.84	7.18	8.22	7.94	2.38	5.38
8	Female	584.26	6.92	7.78	7.67	2.28	5.18
9	Male	757.04	7.73	8.80	8.52	2.47	5.71
9	Female	678.47	7.36	8.28	8.14	2.35	5.49
10	Male	886.57	8.21	9.39	9.00	2.57	6.00
10	Female	758.66	7.76	8.74	8.52	2.42	5.76

The several measurements of Table IV show an interesting relation when the tarso-metatarsus is taken into account (Figs. 28-29); but within themselves a quadratic relation is not found (Figs. 30-32).

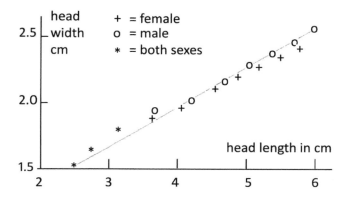

Fig. 30. *Comparison of the growth in head length and head width in a chicken. The straight line is a non-calculated line uniting the smallest and the largest value.*

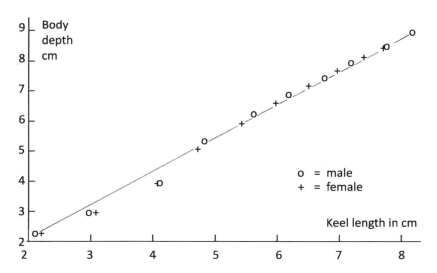

Fig. 31. *Comparison of the growth in keel length and body depth in a chicken. The straight line is a non-calculated line uniting the smallest and the largest value.*

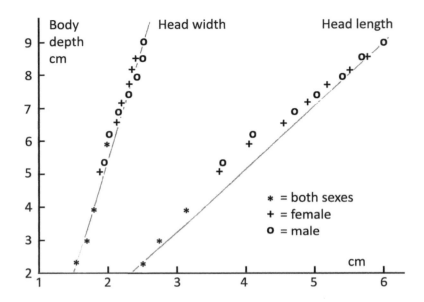

Fig. 32. *Comparison of the growth in head width and head length with body depth. The two straight lines are non-calculated lines uniting the smallest and the largest values.*

Figures 30-32 show four comparisons. In all cases there is distinct relation between the set of measurements that is compared. In three cases the younger stages are found on one side of the straight line (either left or right), therefore, a consistent pattern is not found.

HUMAN GROWTH

D'Arcy Wentworth Thompson considered in his book "On growth and form" (1942) that for "the rate of growth in man" one has to consider the first data published by Quetelet (1835) in his "Essai de Physique Sociale". He wrote: "this epoch-making book is packed with information in regard to human growth and form; and it stands out as the first great essay in which social statistics and organic variation are dealt with from the point of view of the mathematical theory of probabilities". In 1871 Quetelet published all his data on the Belgian population in his book "Anthropométrie". I used a small part of this data gathered from my ancestors dating back almost two centuries. I will leave it to the anthropologists to make the necessary comparisons with more recent data from various parts of the world. My only aim in the following sections is to indicate how our growth stops.

Influence of girth on growth in length

Quetelet (1871) gave three measurements of girth for the same males and females: one at the level of the sternum, one at the waist and one at hip level.

Six equations were obtained.

Male at sternum level: $y = -5.19 \, x^2 + 9.25 \, x - 2.42$

Female at sternum level: $y = -8.35 \, x^2 + 13.18 \, x - 3.55$

Male at waist level: $y = -9.11 \, x^2 + 14.64 \, x - 4.19$

Female at waist level: $y = -22.65 \, x^2 + 30.024 \, x - 8.387$

Male at hip level: $y = -4.30 \, x^2 + 8.45 \, x - 2.32$

Female at hip level: $y = -7.72 \, x^2 + 12.33 \, x - 3.36$

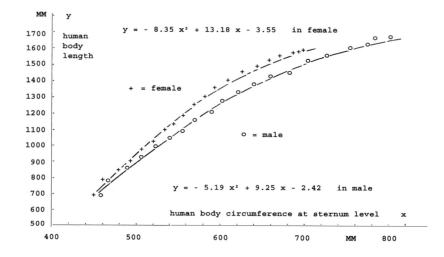

Fig. 33. *Comparison of human body circumference at sternum level during growth from age 1 to 20. The very similar measurements at age 1 suggest that young girls present a much larger negative a-value (-8.35) than young boys (-5.19). Therefore, girls have a greater influence on the diminution in the rate of body growth than boys. For each additional 1 cm growth in body circumference at sternum level, there is a diminution of 0.167 cm for girls and 0.1039 cm for boys. (courtesy of* Annals of Human Biology).

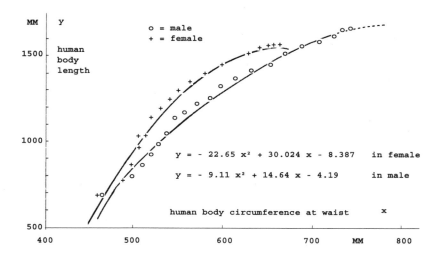

Fig. 34. *Comparison of human body circumference at waist level during growth from age 1 to 20. The very similar measurements at the age 1 suggest that young girls present a much larger negative a-value (-22.65) than young boys (-9.11). Therefore, girls have a greater influence on the diminution in the rate of body growth than boys. For each additional 1 cm growth in body circumference at waist level, there is a diminution of 0.453 cm for girls and 0.1822 cm for boys. (courtesy of* Annals of Human Biology).

This very high negative a-value in young females strongly hinders body growth, as can be seen in Table V, which shows the influence of the growth in girth at waist level on body growth. The theoretical value of the corresponding body length calculated according to the formula obtained is added. In the third column one can observe that the additional growth in body length diminishes even to the point of becoming negative at a waist circumference of 0.67 m. In the fourth column the constant difference is found by which the growth in length diminishes. This number 0.453 is twice 0.2265, i.e. the a-value of the corresponding equation (the equation gives the relation in metres, whereas in Table V the calculation is in cm).

TABLE V

In the first column the theoretical units of the body circumference at waist level in females are given. In the second column the calculated Y-values for the units of X are given. Column three shows the differences between the calculated values per cm increase in body circumference. The last column shows the constant decrease of the increase. At a circumference of 67 cm the additional growth in body length becomes theoretically negative, i.e. growth practically stops.

Body circumference at waist level in centimetre	Corresponding body length in centimetre	Additional growth in body length per cm growth in waist	Constant second difference in centimetre
61	148.4935		
		+ 2.1405	
62	150.634		- 0.453
		+ 1.6875	
63	152.3215		- 0.453
		+ 1.2345	
64	153.556		- 0.453
		+ 0.7815	
65	154.3375		- 0.453
		+ 0.3285	
66	154.666		- 0.453
		- 0.1245	
67	154.5415		

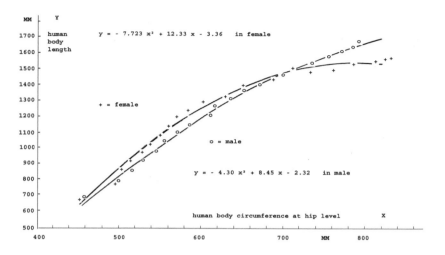

Fig.35. *Comparison of human body circumference at hip level during growth from age 1 to 20. The very similar measurements at age 1 suggest that young girls present a higher negative a-value (-7.723) than young boys (-4.30). Therefore, girls have a greater influence on the diminution in the rate of body growth than boys. For each additional 1 cm growth in body circumference at hip level, there is a diminution of the increase in body length of 0.1545 cm for girls and 0.086 cm for boys. The difference in growth pattern is small; the two curves cross each other showing that at age 14-15 boys and girls reach a similar relation between hip level circumference and body length. (courtesy of* Annals of Human Biology).*

The three studies on the relation between body circumference and body length indicate that growth in body length is negatively influenced by growth in circumference. The largest influence is seen at the waist level in girls, which results in a shorter body length compared to boys.

TABLE VI

Human growth from birth to age 20 according to Quetelet (1871).

AGES	BODY LENGTH IN MM		CIRCUMFERENCE AT STERNUM LEVEL IN MM		CIRCUMFERENCE AT HIP LEVEL IN MM		CIRCUMFERENCE AT WAIST LEVEL IN MM	
	MALES	FEMALES	MALES	FEMALES	MALES	FEMALES	MALES	FEMALES
1 DAY	500	494	302	297	242	240	281	278
1 YEAR	698	690	458	450	458	452	465	459
2 YEAR	791	781	467	460	500	492	498	487
3 YEAR	864	854	489	478	517	507	510	493
4 YEAR	927	915	505	492	529	516	518	500
5 YEAR	987	974	523	507	543	530	527	503
6 YEAR	1046	1031	539	520	555	540	535	507
7 YEAR	1104	1087	555	533	571	549	547	512
8 YEAR	1162	1142	570	544	585	558	558	520
9 YEAR	1218	1196	588	556	611	568	571	530
10 YR	1273	1249	602	568	617	584	585	540
11 YR	1325	1301	621	581	635	603	597	551
12 YR	1375	1352	641	594	652	628	613	563
13 YR	1423	1400	660	610	671	655	632	580
14 YR	1469	1446	682	626	692	688	653	597
15 YR	1513	1488	704	643	713	729	670	613
16 YR	1554	1521	728	658	734	760	688	628
17 YR	1594	1546	756	672	753	787	707	640
18 YR	1630	1563	777	684	774	810	722	649
19 YR	1655	1570	798	693	783	823	733	655
20 YR	1669	1574	813	700	790	828	741	661

Influence of growth in length on head growth

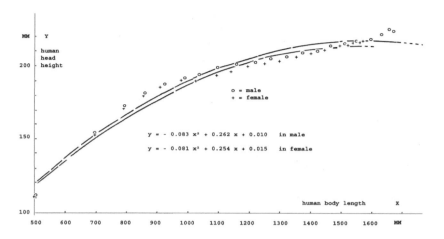

Fig.36. *Comparison of head growth in relation to body length in boys and girls from new-born to age 20. The calculated curves give the relationship between both measurements in metres. (courtesy of* Annals of Human Biology).

The growth pattern in both sexes is very similar and so is the calculated a-value. The growth in body length negatively influences the growth in height of the head.

For every 10 cm increase in body length the head growth diminishes constantly by 0.162 cm in girls and 0.166 cm in boys.

AGES	BODY LENGTH IN MM		HEAD LENGTH IN MM		DISTANCE BETWEEN NIPPLES	
	MALES	FEMALES	MALES	FEMALES	MALES	FEMALES
1 DAY	500	494	111	111	70	70
1 YEAR	698	690	154	154	107	106
2 YEAR	791	781	175	172	111	110
3 YEAR	864	854	182	180	115	114
4 YEAR	927	915	188	184	118	117
5 YEAR	987	974	192	188	121	119
6 YEAR	1046	1031	193	190	125	120
7 YEAR	1104	1087	198	193	130	123
8 YEAR	1162	1142	201	196	135	125
9 YEAR	1218	1196	203	199	140	128
10 YR	1273	1249	205	201	145	132
11 YR	1325	1301	207	203	150	136
12 YR	1375	1352	209	206	155	142
13 YR	1423	1400	211	208	159	148
14 YR	1469	1446	213	210	164	155
15 YR	1513	1488	215	213	169	163
16 YR	1554	1521	217	215	174	175
17 YR	1594	1546	219	217	180	187
18 YR	1630	1563	222	218	186	197
19 YR	1655	1570	223	219	189	201
20 YR	1669	1574	227	220	192	203

TABLE VII

Human growth from birth to age 20 according to Quetelet (1871).

Body weight during growth

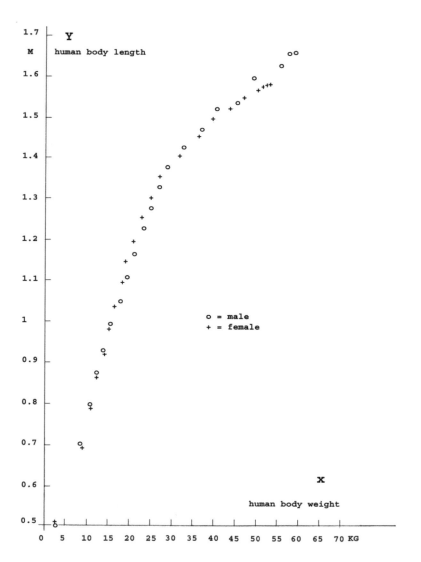

Fig. 37. *Relationship between body length and body weight in boys and girls from new-born to age 20.*

Quetelet used two sets of measurements, one of which was probably obtained in 1835 and the other, in 1840 (Table VIII). He calculated the mean of these two sets, however, between the two sets there is quite some variation and therefore, the mean calculated by Quetelet and used in Fig. 37 is not considered to be accurate.

The diagram shows that during growth both sexes display a similar increase in weight, but as males have a longer growth period, they also become heavier. When growth in height stops, only body weight continues to increase in both sexes. In contrast to the previous body measurements, weight is highly variable and can be influenced; therefore, a parabolic curve is inappropriate. As the specific gravity of men is slightly more than 1, weight in kilograms is comparable to volume in litres.

TABLE VIII

Table in Quetelet showing the results on weight in two Belgian populations (in French)

ANTHROPOMÉTRIE.

AGE.	POIDS DU CORPS DE L'HOMME.				DE LA FEMME.			
	1835	1840	Moyenne.	Accroiss¹ annuel.	1835	1840	Moyenne.	Accroiss¹ annuel.
Naissance. .	3ᵏ2	3ᵏ0	3ᵏ1	.	2ᵏ9	3ᵏ0	3ᵏ0	
0 à 1 an. .	9 4	8 6	9 0	5ᵏ9	8 8	8 4	8 6	5ᵏ6
2 ans. .	11 0	11 0	11 0	2 0	10 7	11 3	11 0	2 4
3 " .	12 4	12 6	12 5	1 5	11 8	13 0	12 4	1 4
4 " .	14 2	13 8	14 0	1 5	13 0	14 8	13 9	1 5
5 " .	15 8	16 0	15 9	1 9	14 4	16 2	15 3	1 4
6 " .	17 2	18 5	17 8	1 9	16 0	17 6	16 7	1 4
7 " .	19 1	20 7	19 7	1 9	17 5	18 1	17 8	1 1
8 " .	20 8	22 5	21 6	1 9	19 1	18 6	19 0	1 2
9 " .	22 6	24 4	23 5	1 9	21 4	20 6	21 0	2 0
10 " .	24 5	25 9	25 2	1 7	23 5	22 7	23 1	2 1
11 " .	27 1	27 0	27 0	1 8	25 6	25 4	25 5	2 4
12 " .	29 8	28 2	29 0	2 0	29 8	28 2	29 0	3 5
13 " .	34 4	31 9	33 1	4 1	32 9	32 1	32 5	3 5
14 " .	38 8	35 5	37 1	4 0	36 7	36 0	36 3	3 8
15 " .	43 6	38 7	41 2	4 1	40 4	39 5	40 0	3 7
16 " .	46 7	44 1	45 4	4 2	43 6	42 9	43 5	3 5
17 " .	52 8	46 6	49 7	4 3	47 3	47 0	46 8	3 3
18 " .	55 8	52 0	53 9	4 2	49 0	51 0	49 8	3 0
19 " .	58 0	57 2	57 6	3 7	51 6	53 5	52 1	2 3
20 " .	60 1	59 0	59 5	1 9	52 3	55 0	53 2	1 1
21 " .	61 2	61 2	61 2	1 7	52 4	56 1	54 3	0 8
22 " .	61 4	64 4	62 9	1 7	52 5	57 0	54 8	0 5
23 " .	61 5	67 5	64 5	1 6	52 8	57 7	55 2	0 4
25 " .	62 9	69 6	66 2	1 7	53 3	56 2	54 8	0 3
27 " .	63 3	68 4	65 9		53 8	56 5	55 1	0 1
30 " .	63 7	68 6	66 1		54 3	56 4	55 3	0 0

Nipple distance

A comparison of the distance between the nipples and body circumference at the sternum yielded interesting findings. All the measurements of the males and females below age 15 together form one almost straight line that represents more or less 24%. There is an almost constant relation between nipple distance and body circumference at the sternum from birth to adulthood for males. For females the distance between the nipples increases from the age of 15.

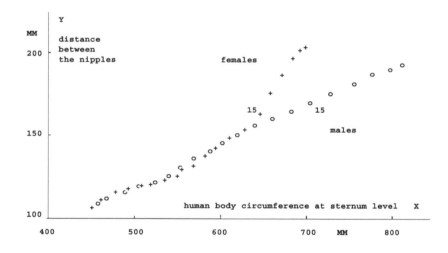

Fig. 38. *Comparison between body circumference at sternum level and nipple distance. The number 15 refers to age. (courtesy of* Annals of Human Biology).

Relationship between arm span and body length

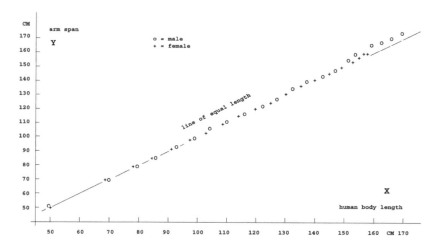

Fig. 39. *Comparison between arm span and body length in the Belgian population measured by Quetelet in the XIXth century.*

The measurements of arm span versus body length in the Belgian population recorded by Quetelet at the beginning of the XIXth century show that from new-born to adult both sexes present a constant relation in which both measurements are the same until age 16. Thereafter, males show additional growth in arm span. This well-known relationship was recorded by Leonardo da Vinci in his drawing of the Vitruvian man (Fig. 40). However, the arm span to body length ratio can vary; for example, it was found to be significantly different in two ethnic groups, namely in both sexes of Afro-Caribbean's and Asian males (Reeves *et al.*, 1996).

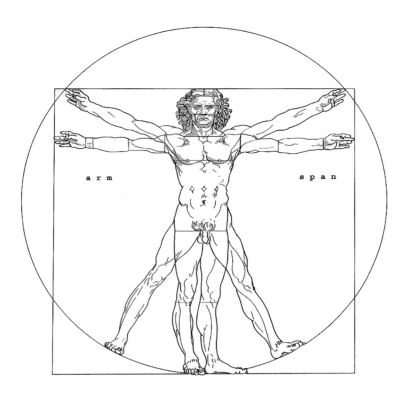

Fig. 40. *Leonardo da Vinci's drawing of the Vitruvian man shows an adult male with a slightly longer arm span compared to his height.*

Conclusion

Although Quetelet's tables summarise only population studies and not growth studies, it is apparent that also here a quadratic parabola closely follows the observations. Therefore, one can understand how growth is regulated and even that this occurs very precisely from birth onwards.

In humans several growth factors regulate the relationships between some of the growth axes in the following ways. First, growth in body circumference negatively influences growth in body length so that eventually, growth in body length stops. In girls the negative influence is greater resulting in shorter body length and larger body circumference, especially at waist level (the development of the pelvis is necessary for reproduction). Second, growth in body length negatively influences head growth. Third, during growth some relationships remain constant.

GENERAL DISCUSSION

By using the formula

$$y = a\,x^2 + b\,x + c$$

In the study of differential growth, I assume that measurement y is the dependent variable and that x is the independent variable. But the examples given show that growth in general is more complicated. Moreover, the given measurements are probably not taken along the important growth axes. Nevertheless, the negative sign of the quadratic factor "a" in various comparisons indicates why growth stops, namely because the growth of a body part is constantly and negatively influenced by the growth of another body part. This does not happen suddenly or after some growth has occurred; it starts from the very beginning. Therefore, I hypothesise that when growth starts, a very precise growth pattern ensues that cannot be changed.

Growth studies often investigate variations within populations. The results are often not as accurate as those obtained from growth studies that follow one population from birth until adulthood. In the studies on the growth of the rat skull and the chicken all measurements followed very closely the theoretical curve. The study of growth in nematodes revealed that it was even possible to explain why during growth some changes in the morphology occurred.

Body width and body length determine body volume. In nematodes their influence is complex with changes in morphology when necessary, but in most animals (including humans) it is obvious that both items are intimately linked – growth in width negatively influences growth in length/height.

Growth is induced by growth- promoting substances, and influenced by growth- inhibiting substances. The quadratic factor in the quadratic equation could indicate the constantly changing relation between the production of growth- promoting and growth- inhibiting substances. The negative sign of this factor could indicate that the growth- inhibiting substances increase in significance and are finally operate at the same level as the growth- promoting substances. As a consequence, growth ceases. It is also hypothesised that it is not because growth has ceased that the production of both substances ceases; the extrapolation of the obtained curves suggests that growth-inhibiting substances become now constantly more important. If that is so, then the curve characterises what is happening after growth: deterioration, malfunction and finally death. According to this speculation, the same metabolic processes that start when growth starts eventually cause death.

Remarks:

1. What is presented here is a hypothesis that explains some morphometric observations; I do not know of any experimental evidence that backs up this suggestion, neither can I suggest how experimentalists should test these ideas.

2. The present study is built on the interpretation of the quadratic factor alone and despite the high R^2 values it is not known if the quadratic function is the best choice from a statistical point of view.

3. Conclusions drawn from the fitted quadratic functions can only be valid for the domain that is investigated (i.e. the range of dependent variables).

4. Each curve with a negative quadratic factor will decrease again after reaching a maximum. This was only observed in the case of the cribriform plate of the rat skull with disappearance of that bone. Another possibility is that the dependent variable does not stop but continues to fluctuate when the theoretical maximum is reached (e.g. weight vs. length). In most of the observed curves the calculated maximum marked the end of a particular growth process.

5. The choice of which variables were dependent and which were independent was not hypothesis driven. If we were to reverse the variables in the abscissa and ordinate it would produce meaningless results. It is even possible that the equations provided

here are descriptive for the development of shape an organism may take, but do not imply any causality.

6. An analysis based on population averages implies the rather crude assumption that people with an average girth, for example, also have an average height throughout childhood and adolescence.

Growth is not simply time dependent; it depends also on food and metabolism, and these processes need time.

To return to Stack's questions mentioned in the Preface, we could answer thus:

(1) *What controls the absolute size of the whole, or why are we bigger than mice?*

The absolute size of the whole is controlled by the constantly changing equilibrium between the various growth-promoting and growth-inhibiting substances. In the course of evolution the changes in that equilibrium could augment producing bigger men or mice.

(2) *Within a whole, what maintains the constancy of proportions of individual parts?*

The constancy of proportions of individual parts is maintained by an active process that controls the exact amount of the substances that control growth.

(3) *How is a possible change of relative proportions (allometry) produced?*

Allometry is produced by the constant second differences in one growth parameter in relation to another growth parameter. It is not that various parts of an organism grow at a constant relative rate (as mentioned by Bartoletti *et al.*, 1999) but at a constantly changing relative rate.

Presumptions:

1. The use of a quadratic equation in a case of intra-specific allometry could allow us to predict the shape of larger and/or smaller specimens not yet observed.

2. It can be assumed that changes in the growth factor(s) do sometimes occur and have occurred in the past. Such shifts could explain some evolutionary changes.

3. When growth starts it apparently "knows" when it shall stop. Calculations of quadratic curves based on measurements in the beginning of the growth process will probably have a predictive value.

4. In this book only "normal" growth patterns have been studied. The growth factors that are obtained for whatever animal can help to indicate, even in the youngest stages, a possible abnormal growth.

GENERAL CONCLUSIONS

1. Growth has to be considered as a uniformly decelerated movement rather than as an exponential process.

2. Allometry can best be described by a quadratic curve resulting in a parabola (also called a polynomial equation of the second degree). The quadratic factor gives the exact change in growth of the dependent variable y for each unit of growth of the independent variable x. At the apex of the parabola the growth of the dependent variable stops (in most cases) and can be followed by negative growth (exceptionally).

3. Growth, adulthood, ageing and death could be aspects of exactly the same process that starts from the very beginning of multicellularity. Growth is influenced by specific growth-promoting and growth-inhibiting substances. It is hypothesised that growth-inhibiting substances increase in significance so that eventually, growth stops, deterioration ensues and death follows.

ARTICLES ON GROWTH BY THE AUTHOR

GERAERT, E. (1978). On growth and form in Nematoda: oesophagus and body-width in Tylenchida. *Nematologica* 24, 137-158.

GERAERT, E. (1978). On growth and form in nematodes: II. Oesophagus and body width in Dorylaimida. *Nematologica* 24, 347-360.

GERAERT, E. (1979). Growth and form in nematodes: III. Comparison of oesophagus and body shape. *Nematologica* 25, 1-21.

GERAERT, E. (1979). Growth and form in nematodes: IV. Tail length and vulva position. *Nematologica* 25, 439-444.

GERAERT, E. (1980). Differential growth formulae. *Biologisch Jaarboek Dodonaea* 47, 87-95.

GERAERT, E. (2004). Constant and continuous growth reduction as a possible cause of ageing. *International Journal of Developmental Biology* 48, 271-274.

GERAERT, E. (2013). Remarks on the article of Packard (2012) "Julian Huxley, *Uca pugnax* and the allometric method". *Journal of Experimental Biology* 216 (3), 535.

GERAERT, E. (2016). A quadratic approach to allometry yields promising results for the study of growth. *Belgian Journal of Zoology* 146 (1), 14-20.

GERAERT, E. (2018). Differential human growth restudied. *Annals of Human Biology* 45 (2), 110-115.

REFERENCES

BARTOLETTI, S., FLURY, B.D. & NEL, D.G. (1999). Allometric Extension. *Biometrics* 55, 1210-1214.

BARTON, A.D & LAIRD, A.K. (1969). Analysis of allometric and non-allometric differential growth. *Growth* 33, 1-16.

BASIR, M.A. (1950). The morphology and development of the sheep nematode, *Strongyloides papillosus* (Wedl, 1956). *Canadian Journal of Research* 28, 173-196.

BATSCHELET, E. (1975). *Introduction to Mathematics for Life Scientists*. 2nd Ed. 643 pp. Springer-Verlag, Berlin, Heidelberg, New York.

CHAMPY, C. (1924). *Sexualité et hormones*. Doin, Paris.

CHAN, K.F. (1952). Life cycle studies on the nematode *Syphacia obvelata*. *American Journal of hygiene* 56, 14-21.

CHOI, Y.E., GERAERT, E & BAEK, H.S. (1997). Taxonomic study of Dorylaimoidea (Nematoda: Dorylaimida) from Korea. *Korean Journal of applied Entomology* 36 (1), 1-29.

CHU, T.C. (1936). Studies on the life history of *Rhabdias fuscovenosa* var. *catanensis* (Rizzo, 1902). *Journal of Parasitology* 22, 140-160.

CHUANG, S.K. (1962). The embryonic and post-embryonic development of *Rhabditis teres* (A. Schneider). *Nematologica* 7, 317-330.

CUZIN-ROUDY, J. & LAVAL, P. (1975). A canonical discriminant analysis of post-embryonic development in *Notonecta maculata* Fabricius (Insecta: Heteroptera). *Growth* 39, 251-280.

D'ARCY W. THOMPSON (1942, 1968). *On Growth and Form*. University Press, Cambridge.

DECRAEMER, W. (2016). *Nematode Morphology*. Academia Press, Ghent, Belgium.

DOUVRES, F.W. (1957). The morphogenesis of the parasitic stages of *Trichostrongylus axei* and *Trichostrongylus colubriformis*, Nematode Parasites of cattle. *Proceedings of the Helminthological Society of Washington* 24, 4-14.

DUDICH, E. (1923). Über die Variation des *Cyclommatus tarandus* Thunberg (Coleoptera, Lucanidae). *Archiv für Naturgeschichte* 89 (1-4), 62-96.

EISENBACK, J.D. (1989). Identification of Meloidogynids. In: *Nematode Identification and expert Systems Technology.* R. Fortuner: ed. Plenum Press, New York.

FORD, E.H.R. & HORN, G. (1959). Some problems in the evaluation of differential growth in the rat skull. *Growth* 23, 191-204.

GAYON, J. (2000). History of the concept of allometry. *American Zoologist* 40, 748-758.

GOLDBERG, A. (1951). Life history of *Oesophagostomum venulosum*, a nematode parasite of sheep and goats. *Proceedings of the Helminthological Society of Washington* 18, 36-47.

GOULD, S.J. (1966). Allometry and size in ontogeny and phylogeny. *Biological Reviews of the Cambridge Philosophical Society* 41, 587-640.

GOULD, S.J. (1971). Geometric similarity in allometric growth: a contribution to the problem of scaling in the evolution of size. *The American Naturalist* 105 n° 942, 113-136.

GYSELS, H. & VAN DER HAEGEN, W. (1962). Post embryonale ontwikkeling en vervellingen van de vrijlevende nematode *Panagrellus silusiae* (de Man, 1913) Goodey, 1945. *Natuurwetenschappelijk Tijdschrift* 44, 3-20.

HECHLER, H.C. (1968). Postembryonic development and reproduction in *Diploscapter coronata* (Nematoda: Rhabditidae). *Proceedings of the Helminthological Society of Washington* 35, 24-30.

HECHLER, H.C. (1970). Reproduction, chromosome number and post-embryonic development of *Panagrellus redivivus* (Nematoda: Cephalobidae). *Journal of Nematology* 2, 355-361.

HUXLEY, J.S. (1924). Constant differential growth-ratios and their significance. *Nature* 114, 895-896.

HUXLEY, J.S. (1927). Further work on heterogonic growth. *Biologisches Zentralblatt* 47, 151-163.

HUXLEY, J.S. (1932). *Problems of Relative Growth.* Methuen & Co, London.

HUXLEY, J.S. & RICHARDS, O.W. (1931). Growth of the abdomen and the carapace of the Shore-crab *Carcinus maenas*. *Journal of the Marine Biological Association of the U.K.* 17 (03), 1001-1015.

HUXLEY, J.S. & TEISSIER, G. (1936). Terminology of relative growth. *Nature* 137, 780-781.

KAREGAR, A., De LEY, P. & GERAERT, E (1997). A detailed morphological study of *Acromoldavicus skrjabini* (Nesterov & Lisetskaya, 1965) Nesterov, 1970 (Nematoda: Cephaloboidea) from Iran and Spain. *Fundamental and applied Nematology* 20 (3), 277-283.

KIDWELL, J.F. & HOWARD, A (1970). The inheritance of growth and form in the mouse. III. Orthogonal polynomials. *Growth* 34, 87-97.

KIDWELL, J.F. & WILLIAMS, E. (1956). Allometric growth of the dark Cornish Fowl. *Growth* 20, 275-293.

KNELL, R.J., POMFRET, J.C. & TOMKINS, J.L. (2004). The limits of elaboration: curved allometries reveal the constraints on mandible size in stag beetles. Proceedings of the Royal Society of London. Series B. *Biological Sciences* 271, 523-528.

LEVINTON, J. (1988). *Genetics, paleontology and macroevolution.* Cambridge University Press, Cambridge.

LUDWIG, H. (1938). Die Variabilität von *Rhabditis teres* unter veränderten Ernährungsbedingungen. *Zeitschrift fur wissenschaftlichen Zoologie* 131, 291-356.

MARTIN, L. (1960). Homométrie, allométrie et cograduation en biométrie générale. *Biometrische Zeitschrift* 2, 73-97.

MATOFF, K. (1936). Beobachtungen über die larval Entwicklung von *Strongyloides papillosus* (Wedl, 1856) und Infektionsversuche mit filariformen Larven. *Zeitschrift für Parasitenkunde* 8, 474-491.

MOORTHY, V.N. (1938).Observations on the life history of *Camallanus sweeti. Journal of Parasitology* 24, 323-342.

NEEDHAM, A.E. (1950). The form-transformation of the abdomen of the female pea-crab, *Pinnotheres pisum* Leach. *Proceedings of the Royal Society of London* B, 115-136.

NEEDHAM, A.E. (1957). The Quadratic Relation in Differential Growth. *Nature* 180, 1293.

NEEDHAM, A.E. (1964). *The Growth Process in Animals*. Sir Isaac Pitman & Sons LTD, London; 522 pp.

PACKARD, G.C. (2012). Julian Huxley, *Uca pugnax* and the allometric method. *Journal of Experimental Biology* 215, 569-573.

QUETELET, A. (1835). *Sur l'Homme et le Développement de ses Facultés, ou Essai de Physique Sociale*. Tome second. Bachelier, Paris.

QUETELET, A. (1871). *Anthropométrie ou Mésure des Différentes Facultés de L'Homme*. Muquardt, Brussels, Ghent, Leipzig.

REEVES, S.L., VARAKAMIN, C. & JEYA HENRY (1996). The relationship between arm-span measurement and height with special reference to gender and ethnicity. *European Journal of Clinical Nutrition* 50 (6), 398-400.

RIGBY, M.C. & RIGBY, E. (2014). 7.20. Order Camallanida: Superfamilies Anguillicoloidea and Camallanoidea. In: A. Schmidt-Rhaesa (Ed.): *Handbook of Zoology: Nematoda*, pp. 637-659.

SHIH, HSI-TE; NG, PETER K.L.; DAVIE, PETER J.F. , et al. (2016). Systematics of the family Ocypodidae Rafinesque, 1815 (Crustacea: Brachyura), based on phylogenetic relationships, with a reorganization of subfamily rankings and a review of the taxonomic status of *Uca* Leach, 1814, sensu lato and its subgenera. *Raffles Bulletin of Zoology* 64, 139-171.

SHINGLETON, A. (2010). Allometry: The Study of Biological Scaling. *Nature Education Knowledge* 1 (9), 2.

SLACK, J.M.W. (1999). Problems of Development; The Microcosm and the Macrocosm. In: *On Growth and Form: Spatio-temporal Pattern Formation in Biology*. Eds. Chaplain, M.A.J., Singh, G.D. & McLachlan, J.C. New York, John Wiley & Son, pp. 1-12.

SMITH, S.I. (1870). Notes on American Crustacea. Number I. Ocypodoidea. *Transactions of the Connecticut Academy of Arts and Science* 2, 113-176. Plates 2-4.

SNELL, O. (1891). Die Abhängigkeit des Hirngewichtes von dem Körpergewicht und der geistigen Fâhigkeiten. *Archiv für Psychologische und Nervenkrankheiten* 23, 436-446.

STERN, D.L. & EMLEN, D.J. (1999). The Developmental Basis for Allometry in Insects. *Development* 126, 1091-1101.

TEISSIER, G. (1931). Recherches morphologiques et physiologiques sur la croissance des insectes. *Travaux de la Station Biologique de Roscoff* 9, 29-238.

THUNBERG, P. (1806). Lucani Monographia, elaborata. *Mémoires de la Sociéte Imperiale des Naturalistes de Moscou* 1, 183-206, 1 pl.

TRIANTAPHYLLOU, A.C. & HIRSCHMANN, H. (1960). Post-infection development of *Meloidogyne incognita* Chitwood, 1949 (Nematoda: Heteroderidae). *Annals of the Institute of Phytopathology*, Benaki 3, 3-11.

TROCCOLI, A. & GERAERT, E. (1995). Some species of Tylenchida (Nematoda) from Papua New Guinea. *Nematologia Mediterranea* 23, 283-298.

WALKER, G.F. & KOWALSKI, C.J. (1971). A two-dimensional coordinate model for the quantification, description, analysis, prediction and simulation of craniofacial growth. *Growth* 35, 191-211.

WEST, G.B., BROWN, J.H. & ENQUIST, B.J. (1997). A General model for the Origin of Allometric Scaling Laws in Biology. *Science* 276, 122-133.

WEST, G.B., BROWN, J.H. & ENQUIST, B.J. (2001). A general model for ontogenetic growth. *Nature* 413, 628-631.

WILLIAMS, P. (2013). *Why do we stop growing?* Available online at: http://www.nbcnews.com/id/3076703/t/why-do-we-stop-growing/

ZEGER, S.L. & HARLOW, S.D. (1987). Mathematical model from laws of growth to tools for biological analysis: fifty years of growth. *Growth* 51, 1-21.

Academia Press
Ampla House
Coupure Rechts 88
9000 Gent
België

www.academiapress.be

Academia Press is a subsidiary of Lannoo Publishers.

ISBN 978 94 014 6214 3
D/2019/45/313
NUR 922

Etienne Geraert
Why we stop growing
Gent, Academia Press, 2019, 96 p.

Lay-out: Keppie & Keppie

© Etienne Geraert
© Lannoo Publishers

*No part of this publication may be reproduced in print, by photocopy, microfilm
or any other means, without the prior written permission of the publisher.*